FLORA OF TROPICAL EAST AFRICA

LENTIBULARIACEAE

P. Taylor

Herbaceous terrestrial, epiphytic or aquatic plants all with specialized organs for the capture of small organisms. Roots frequently absent. Leaves rosulate or scattered on stolons, entire or divided, sometimes polymorphic. Inflorescence terminal or lateral, peduncled, racemose, simple or sparingly branched, bracteate; lowest bracts (scales) usually barren; bracteoles 2 or 0 or ± connate with the bract, usually at the base of the pedicels. Flowers hermaphrodite, zygomorphic. Calyx deeply 2–4- or 5-partite, almost regular or ± 2-lobed or the sepals free to the base, persistent and often accrescent. Corolla gamopetalous, 2-lipped, usually spurred, rarely saccate, usually violet or yellow; tube very short; upper lip interior, entire or ± 2- or rarely more lobed; lower lip entire or 2–5-lobed, usually with a raised ± gibbous palate. Stamens 2, anticous, inserted at the base of the corolla; filaments usually short, usually curved, rarely longer and geniculate; anthers 2-thecous; thecae diverging, ± confluent, dehiscing by a common slit. Ovary superior, 1-locular; carpels 2, median; style simple, usually short or very short, rarely longer and geniculate; stigma ± 2-lipped, the upper lip smaller than the lower or ± obsolete; placenta free central or free basal, usually ovoid or globose; ovules usually numerous, sessile, rarely fewer or 2, anatropous. Fruit a 1-celled capsule, 2–many-seeded, dehiscing by longitudinal slits or by pores or circumscissile or rarely indehiscent. Seeds small or very small, very variously shaped; testa thin or spongy or corky, rarely mucilaginous; endosperm 0; embryo undifferentiated.

Four genera with about 200 species mostly in the tropics and the temperate northern hemisphere.

Calyx-lobes 2; leaves very varied; traps small, globose or
    ovoid with various oral appendages, the mouth with a
    sensitive door which opens and closes to capture small
    organisms; capsule dehiscing by slits or pores, circum-
    scissile or indehiscent .   .   .   .   .   .   1. **Utricularia**
Calyx-lobes 5; leaves ± spathulate, rosulate; traps passive,
    stalked, tubular from an ellipsoid base and terminating
    in 2 twisted arms; capsule multiple-circumscissile .   2. **Genlisea**

1. UTRICULARIA

L., Sp. Pl.: 18 (1753) & Gen. Pl., ed. 5: 11 (1754); Stapf in F.T.A. 4 (2): 469 (1906); P. Taylor in K.B. 18: 1 (1964) & in Fl. Afr. Centr., Lentib.: 2 (1972)

Herbs, annual or perennial, terrestrial, epiphytic or aquatic. Vegetative parts not clearly differentiated but consisting of stems modified to function as roots, stems, leaves and specialized organs (traps) for the capture of small organisms. Root-like organs (rhizoids) usually descending from the base of

1

the inflorescence, usually filiform. Stem-like organs (stolons) arising with the rhizoids at the inflorescence base; in the terrestrial and epiphytic species usually short and delicate but sometimes developing into fleshy tubers; in the aquatic species usually more robust and longer. Foliar organs (leaves) either rosulate at the inflorescence base or alternate, opposite or verticillate on the stolons; in the terrestrial and epiphytic species entire, erect or thalloid, capillary, linear, orbicular or peltate; in the aquatic species ± dichotomously divided into capillary segments. Traps hollow, globose or ovoid, usually stalked, with a mouth which may be basal (adjacent to the stalk), lateral or terminal (opposite to the stalk); mouth provided externally with 2 lips which bear very diverse appendages. Inflorescence racemose, bracteate; scape usually simple, sometimes branched, usually filiform, erect or twining, usually glabrous, sometimes papillose, glandular or hairy, provided (especially in the terrestrial and epiphytic species) with sterile bracts (scales); raceme usually elongated, rarely short and subcapitate; pedicels usually short, terete, flattened or ± winged, often reflexed or recurved in fruit; bracts persistent, basifixed or produced below the point of insertion or peltate; bracteoles 2 or absent or sometimes ± fused with the bract, inserted with the bract at the base of the pedicel or rarely with the calyx-lobes at the apex of the pedicel. Calyx-lobes 2, ± equal or sometimes very unequal, usually free, sometimes ± united at the base, persistent and usually accrescent, sometimes very markedly so; upper lobe usually entire, lower lobe usually emarginate or bidentate, rarely both lobes dentate or fimbriate. Corolla bilabiate, glabrous, glandular or pubescent; throat closed or sometimes open; upper lip usually ± erect, limb entire, emarginate or bilobed; lower lip usually larger, spurred or rarely saccate at the base, palate usually raised and gibbous, limb spreading or deflexed, entire, emarginate or ± deeply 2–5-lobed; spur ± parallel to the lower lip or divergent at an acute or obtuse angle or rarely in the same plane. Stamens 2, inserted at the base of the corolla; filaments straight or curved, usually twisted, sometimes winged; anthers dorsifixed, ± ellipsoid, the thecae ± confluent. Ovary globose or ovoid, unilocular; style usually short, often indistinct, persistent; stigma bilabiate, lower lip usually much larger than the upper which may be obscure or obsolete; ovules 2-many, sessile on a ± fleshy basal or free central placenta, anatropous. Fruit a capsule, ± globose, dehiscing by longitudinal slits or by pores or circumscissile or indehiscent. Seeds 1–many, usually small or very small, globose, ovoid, truncate conical, narrowly cylindrical, fusiform, lenticular or prismatic, smooth, verrucose, reticulate, glochidiate or variously winged. Embryo undifferentiated.

About 150 species, mainly tropical but with a few species in the northern temperate region; 33 species occur in tropical Africa of which 23 are represented in the Flora area.

Terrestrial or epiphytic affixed herbs; leaves entire, often evanescent; bracteoles present or if absent then bracts peltate:

Bracts and bracteoles not produced below the point of insertion:

Bracts more than 4 times as wide as the bracteoles; fruiting pedicels flattened or winged, as long as or longer than the calyx:

Fruiting pedicels reflexed . . . . 5. *U. baouleënsis*

Fruiting pedicels erect or spreading:

Seeds ovoid; testa-cells elongate; capsule-wall locally thickened at line of dehiscence; corolla yellow:

Seeds smooth, with a prominent hilum;
 leaves 1-nerved. . . . .   1. *U. scandens*
Seeds verrucose; leaves 3- or more nerved:
 Corolla less than 10 mm. long, palate
  smooth; leaves rosulate, usually
  present at anthesis . . .   2. *U. andongensis*
 Corolla 15–20 mm. long, palate usually
  4-carinate; leaves not rosulate nor
  usually present at anthesis . .   3. *U. prehensilis*
Seeds globose; testa-cells isodiametric; capsule-wall uniformly membranous; corolla
 usually violet or mauve (rarely yellow)   4. *U. spiralis*
Bracts not more than twice as wide as the bracteoles; fruiting pedicels terete, usually shorter
 than the calyx:
Lower lip of corolla ± deeply 5-lobed . .   6. *U. pentadactyla*
Lower lip of corolla entire or 3-lobed:
 Upper lip of corolla orbicular or broadly ovate,
  broader than the upper calyx-lobe:
 Corolla 4–5 mm. long, yellow, lower lip 3-
  lobed, $\frac{1}{2}$–$\frac{1}{3}$ as long as the spur; inflorescence-axis straight . .   7. *U. firmula*
 Corolla 6–15 mm. long, usually violet and
  yellow (rarely wholly yellow), lower lip
  entire or obscurely 3-crenate, ± $\frac{2}{3}$ as
  long as the spur; inflorescence-axis
  flexuous . . . .   8. *U. welwitschii*
 Upper lip of corolla narrowly oblong or
  oblong-obovate, narrower than the
  upper calyx-lobe:
 Lower lip of corolla ± $\frac{1}{2}$ as long as the spur;
  corolla 3–9 mm. long; calyx not pli-
  cate; base of scape usually papillose .   9. *U. arenaria*
 Lower lip of corolla $\frac{2}{3}$ to equal or longer
  than the spur; corolla 6–15 mm. long;
  calyx ± plicate; base of scape always
  glabrous . . . .   10. *U. livida*
Bracts and, where present, bracteoles produced
 below the point of insertion:
Bracteoles present:
 Calyx-lobes ± equal; bracts and bracteoles
  very shortly produced below the point of
  insertion; leaves peltate . .   11. *U. pubescens*
 Calyx-lobes very unequal, the upper much
  larger; bracts and bracteoles quite distinctly produced below the point of
  insertion; leaves reniform or ligulate:
 Upper calyx-lobe orbicular, emarginate;
  lower lip of corolla 5-lobed; seeds
  glochidiate . . . .   12. *U. striatula*
 Upper calyx-lobe broadly rhomboid, not
  emarginate; lower lip of corolla entire
  or obscurely 2-crenate; seeds not
  glochidiate . . . .   13. *U. appendiculata*
Bracteoles absent; bracts ± orbicular, peltate .   14. *U. subulata*

Aquatic ± freely floating herbs; leaves divided into capillary segments, always present at anthesis; bracteoles absent:

Inflorescence provided with a whorl of inflated floats on or at the base of the scape:

Stolons and traps villous; calyx-lobes much shorter than the capsule; spur of corolla inflated, much longer than the lower lip .   15. *U. benjaminiana*

Stolons and traps not villous; calyx-lobes as long as or longer than the capsule; spur of corolla shorter than the lower lip:

Floats cylindrical-fusiform, 6–8 times as long as broad; corolla usually mauve, rarely yellow; scales at leaf-bases denticulate; fruiting calyx-lobes erect, concealing the capsule; seeds prismatic, ± as wide as long   16. *U. inflexa*

Floats narrowly ellipsoid, 2–4 times as long as wide; corolla yellow; scales at leaf-bases divided into hispid linear segments; fruiting calyx-lobes reflexed exposing the capsule; seeds 2–3 times as wide as long.   17. *U. stellaris*

Inflorescence not provided with a whorl of floats:

Corolla distinctly spurred, yellow:

Traps inserted in the angles of the bifurcations of the leaves   .   .   .   .   .   18. *U. reflexa*

Traps inserted laterally on the leaf-segments:

Stolons flattened; internodes 3–10 cm.; flowers numerous; fruiting pedicel recurved; seeds with a narrow entire wing   .   .   .   .   .   .   19. *U. foliosa*

Stolons terete; internodes up to 2 cm.; flowers few; fruiting pedicels not recurved:

Leaves repeatedly divided into very numerous segments; ultimate segments setulose; ripe fruits never produced   .   .   .   .   .   20. *U. australis*

Leaves 2–3 times forked only; ultimate segments not setulose; fruits abundantly produced; seeds with a broad irregular wing   .   .   .   .   21. *U. gibba*

Corolla saccate, white or cream   .   .   .   22. *U. cymbantha*

---

1. **U. scandens** *Benj.* in Linnaea 20: 309 (1847); P. Taylor in F.W.T.A., ed. 2, 2: 378 (1963) & in K.B. 18: 46 (1964) & in Fl. Afr. Centr., Lentib.: 7, t. 1 (1972). Type: India, Madras, Arcot, *Chuter* (K, holo.!)

Terrestrial herb. Rhizoids and stolons capillary, few from the base of the scape. Leaves few, branch-opposed on the stolons, linear, 1-nerved, up to 10 mm. long and 1 mm. wide. Traps scattered on the stolons and leaves, globose, shortly stalked, glandular, 0·6–1·0 mm. long; mouth basal; upper lip with 2 simple subulate appendages; lower lip with a single shorter truncate or shortly bifid appendage. Inflorescence erect or twining, 3–35 cm. high; scape filiform, glabrous; flowers 1–8, distant, usually alternating with sterile

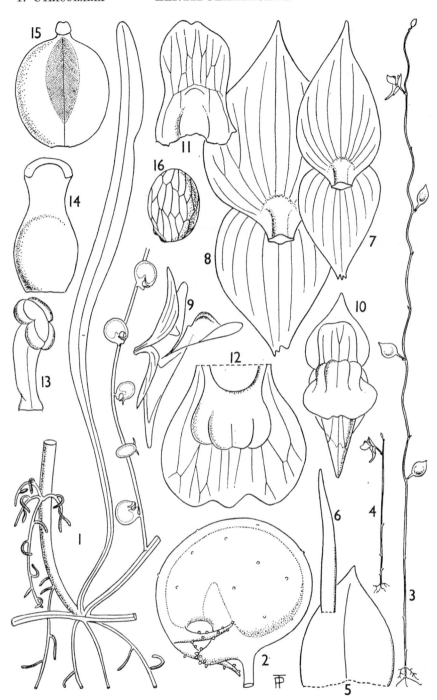

FIG. 1. *UTRICULARIA SCANDENS* subsp. *SCANDENS*—**1,** base of flowering plant showing scape-base, rhizoids, stolons, leaves and traps, × 8; **2,** trap, × 60; **3,** twining inflorescence, × 1; **4,** non twining inflorescence, × 1; **5,** bract, × 30; **6,** bracteole, × 30; **7,** calyx from flower, × 12; **8,** calyx from fruit, × 12; **9,** flower, lateral view, × 8; **10,** flower, abaxial view, × 8; **11,** upper lip of corolla, × 12; **12,** lower lip of corolla, × 12; **13,** stamen, × 30; **14,** pistil, × 30; **15,** capsule, × 12; **16,** seed, × 90; 1–3, 15, 16, from *Milne-Redhead & Taylor* 8008b; 4, 9, 12, from *Milne-Redhead & Taylor* 8008d; 5–8, 10, 11, 13, 14, from *Milne-Redhead & Taylor* 8008.

bracts; scales few, similar to the bracts; bracts basifixed, broadly ovate-deltoid, acute or acuminate, nerveless, 1–1·5 mm. long; bracteoles linear-lanceolate, ± 1 mm. long; pedicels spreading or erect, 1–3 times as long as the flowering calyx, narrowly winged. Calyx-lobes broadly ovate, 2·5–3 mm. long at anthesis, up to 5 mm. long in fruit, decurrent, slightly unequal; upper longer, acute or acuminate; lower shortly 2–3-dentate. Corolla yellow, 5–15 mm. long; upper lip shorter than to about twice as long as the upper calyx-lobe, oblong to orbicular, apex ± emarginate; lower lip orbicular, apex rounded, entire or 2–3-crenate; palate raised, usually 2–4-gibbous; spur subulate, acute, ± curved. Filaments linear, ± straight; anther-thecae ± confluent. Ovary ovoid; style short; stigma-lips semi-orbicular, the upper smaller. Capsule oblong-ovoid, dorsiventrally compressed, 2–2·5 mm. long, dehiscing by a longitudinal abaxial slit with thickened margins. Seeds ovoid or elliptic, smooth, ± 0·2 mm. long; hilum prominent; testa-cells distinct, considerably longer than broad.

subsp. **scandens**

Pedicels about as long as the calyx; corolla 5–7 mm. long; upper lip shorter and narrower than the upper calyx-lobe.

TANGANYIKA. Kigoma District: near Kigoma, Buhamba, Feb. 1953, *Ross* 1506!; Songea District: Ulamboni valley 11 km. W. of Songea, Dec. 1955, *Milne-Redhead & Taylor* 8008!
DISTR. **T**4, 8; Zaire (Katanga), Mozambique, Zambia, Rhodesia, South Africa (Transvaal), Madagascar and tropical Asia
HAB. Bogs and damp grassland; 780–1000 m.

SYN. *U. gibbsiae* Stapf in F.T.A. 4 (2): 574 (1906). Type: Rhodesia, Victoria Falls *Gibbs* 177 (K, holo.!, BM, iso.!)

NOTE. *U. scandens* subsp. *schweinfurthii* (Stapf) P. Taylor replaces subsp. *scandens* in northern and western tropical Africa. It has longer pedicels, larger flowers with a longer, relatively broader upper corolla-lip.

2. **U. andongensis** *Hiern*, Cat. Afr. Pl. Welw. 1: 787 (1900); Stapf in F.T.A. 4 (2): 481 (1906); P. Taylor in F.W.T.A., ed. 2, 2: 377 (1963) & in K.B. 18: 38 (1964) & in Fl. Afr. Centr., Lentib.: 8 (1972). Type: Angola, Cuanza Norte, Pungo Andongo, *Welwitsch* 264 (LISU, holo.!, BM, COI, G, K, P, iso.!)

Terrestrial herb. Rhizoids and stolons capillary, few from the base of the scape. Leaves 3–6, rosulate at the base of the scape and usually present at anthesis, linear, 3-nerved, up to 6 cm. long, 1·5–5 mm. wide. Traps few, globose, shortly stalked, glandular, 0·8–1·1 mm. long; mouth basal; upper lip with 2 subulate appendages. Inflorescence erect or rarely twining, 2–15(–20) cm. high; scape filiform, glabrous; flowers 1–8; scales few, similar to the bracts; bracts basifixed, ovate, acuminate, 1·5–2 mm. long, 1–3-nerved; bracteoles linear-subulate, shorter than the bracts; pedicels erect or ascending, (1–)1·5–3 times as long as the calyx, narrowly winged. Calyx-lobes broadly ovate, 1·5–2·5 mm. long at anthesis, becoming longer and relatively broader in fruit, slightly unequal; upper acute; lower shorter, apex bidentate. Corolla yellow, 4–10 mm. long; upper lip ± as long as the upper calyx-lobe, narrowly oblong, apex rounded, truncate or emarginate; lower lip orbicular, apex entire, emarginate or obscurely 3-crenate; palate scarcely raised, obscurely 2-gibbous; spur conical-subulate, acute, curved. Filaments linear, slightly curved; anther-thecae subdistinct. Ovary narrowly ovoid; stigma subsessile, lower lip truncate or rounded, upper lip smaller, truncate. Capsule broadly oblong, dorsiventrally compressed, smaller than the calyx, dehiscing by a longitudinal abaxial slit with thickened margins in the upper half. Seeds few, ovoid, verrucose, 0·4–0·6 mm. long; testa rather loose and corky, cells irregular, much longer than broad.

UGANDA. Mbale District: Sipi, 31 Aug. 1932, *A. S. Thomas* 458!; Masaka District: Lake Nabugabo, 6 Oct. 1953, *Drummond & Hemsley* 4638!; Mengo District: Bugoye, June 1915, *Dummer* 2623!

TANGANYIKA. Bukoba District: Bukoba, June 1931, *Haarer* 2015! & Ihangiro, 22 Feb. 1892, *Stuhlmann* 3348!

DISTR. U3, 4; T1; from Guinée to Zambia and Angola

HAB. Swamp, *Sphagnum* bogs and wet rocks; 1100–1800 m.

3. **U. prehensilis** *E. Mey.*, Comm. Pl. Afr. Austr.: 282 (1837); Stapf in F.T.A. 4 (2): 480 (1906); P. Taylor in K.B. 18: 53 (1964) & in Fl. Afr. Centr., Lentib.: 10, t. 2 (1972). Type: South Africa, Cape Province, *Drège* 4839 (K, S, iso.!)

Terrestrial herb. Rhizoids and stolons capillary, numerous, fasciculate from the base of the scape. Leaves numerous but usually decayed at anthesis, branch-opposed on the stolons, linear to narrowly oblanceolate, up to 10 cm. long and 3 mm. wide, 1–7-nerved. Traps scattered on the stolons and leaves, globose, shortly stalked, glandular, 0·6–1·5 mm. long; mouth basal; upper lip with 2 simple subulate appendages, lower lip with a single laterally compressed appendage. Inflorescence erect or twining, 3–35 cm. high; scape filiform, glabrous; flowers 1–8, distant; scales few, similar to the bracts; bracts basifixed, ovate-deltoid, 1·5–2 mm. long, up to 5-nerved; bracteoles linear, shorter than the bract; pedicels erect, narrowly winged, as long as or longer than the fruiting calyx. Calyx-lobes ovate, up to 5 mm. long at anthesis and 10 mm. long in fruit, ± unequal, upper usually larger, acute, obtuse or subacute, lower truncate or shortly bidentate. Corolla yellow, 8–20 mm. long; upper lip ± twice as long as the upper calyx-lobe, narrowly oblong to broadly spathulate, apex rounded, truncate or ± emarginate; lower lip ± orbicular, apex entire, bifid or obscurely 4-crenate; palate conspicuously raised, usually prominently longitudinally 4-ridged; spur subulate, acute, ± as long as the lower lip. Filaments linear, curved; anther-thecae subdistinct. Ovary ovoid, dorsiventrally compressed; style short; stigma lower lip rounded, upper lip smaller. Capsule ovoid, up to 5 mm. long, dehiscing by a longitudinal abaxial slit with thickened margins. Seeds numerous, ovoid, 0·6–0·8 mm. long, usually verrucose; testa rather loose and corky, the cells distinct, elongate.

KENYA. Trans-Nzoia District: Hoey's Bridge, July 1965, *Tweedie* 3068!; Uasin Gishu District: Eldoret, Dec. 1947, *Bickford in Bally* 5724!; Elgeyo District: Kaibwibich, Aug. 1968, *Thulin & Tidigs* 181!

TANGANYIKA. Iringa District: Kibengu, 16 Feb. 1962, *Polhill & Paulo* 1507!; Njombe District: Mwakete–Njombe road, 17 Jan. 1957, *Richards* 7879!; Songea District: Ulamboni valley 11 km. W. of Songea, 31 Dec. 1955, *Milne-Redhead & Taylor* 8007!

DISTR. K3; T7, 8; Ethiopia to Rhodesia and Angola, South Africa (Transvaal to eastern Cape Province) and Madagascar

HAB. Bogs, marshes and seasonally flooded ground by lakes and rivers; 1200–2100 m.

4. **U. spiralis** *Sm.* in Rees, Cycl. 37, No. 5 (1819); Stapf in F.T.A. 4 (2): 482 (1906), pro parte; P. Taylor in F.W.T.A., ed. 2, 2: 378 (1963) & in K.B. 18: 58 (1964) & in Fl. Afr. Centr., Lentib.: 12 (1972). Type: Sierra Leone, *Afzelius* (LINN, holo.!, BM, BR, S, UPS, iso.!)

Terrestrial herb. Rhizoids and stolons capillary, numerous from the base of the scape. Leaves few, usually decayed at anthesis, leaf-opposed on the stolons, linear, up to 5 cm. long and 2·5 mm. wide, 1–3-nerved. Traps numerous, globose, shortly stalked, glandular, 0·6–1·0 mm. long; mouth basal; upper lip with 2 simple subulate appendages. Inflorescence erect or usually twining, 5–50 cm. high; scape filiform, glabrous; flowers 1–15, usually distant; scales few, similar to the bracts; bracts basifixed, broadly ovate, ± 2 mm. long; bracteoles linear-lanceolate, usually shorter than the bracts; pedicels erect or ascending, 3–6 mm. long at anthesis, up to 15 mm. long in fruit, flattened or ± narrowly winged, always longer than the fruiting calyx. Calyx-lobes subequal, ovate to narrowly ovate, apex of upper acute, of lower minutely bidentate. Corolla usually violet with a dark blue, greenish, yellow or white spot in the throat, rarely wholly yellow or white, 5–30 mm. long; upper lip narrowly oblong to orbicular; lower lip orbicular; palate raised; spur usually curved, acute or rarely obtuse. Filaments linear; anther-thecae subdistinct; ovary ovoid; style indistinct; stigma-lips short, truncate, subequal. Capsule narrowly ovoid, dehiscing by a longitudinal abaxial slit; capsule-wall of uniform thickness. Seeds numerous, globose, 0·2–0·3 mm. in diameter; testa thin, cells distinct, ± isodiametric.

var. spiralis

Corolla 10–30 mm. long; upper lip longer than the upper calyx-lobe, orbicular.

TANGANYIKA. Buha District: Kaberi Swamp, 10 Aug. 1950, *Bullock* 3116!; Songea District: Mbinga, 24 May 1924, *Zerny* 694! & Hanga Farm ENE. of Songea, 27 June 1956, *Milne-Redhead & Taylor* 10737!
DISTR. T4, 8; Guinée to Zambia and Angola
HAB. Swamps, marshes and bogs; 1000–1500 m.

SYN. *U. baumii* Kam. in E.J. 33: 102 (1902); Stapf in F.T.A. 4 (2): 479 (1906). Type: Angola, *Baum* 307 (BM, BR, COI, HBG, K, M, S, W, Z, iso.!)

var. tortilis (*Oliv.*) P. Taylor in Taxon 12: 294 (1963) & in F.W.T.A., ed. 2, 2: 378 (1963) & in K.B. 18: 62 (1964) & in Fl. Afr. Centr., Lentib.: 13 (1972). Type: Angola, *Welwitsch* 258 (LISU, holo.!, BM, C, COI, G, K, P, iso.!)

Corolla 5–10 mm. long; upper lip not or scarcely longer than the upper calyx-lobe, narrowly oblong.

UGANDA. W. Nile District: Koboko, *Eggeling* 1857!; Acholi District: Gulu, 13 Nov. 1941, *A. S. Thomas* 4016a!; Masaka District, Lake Nabugabo, Aug. 1935, *Chandler* 1294!
KENYA. N. Kavirondo District: near Bungoma, Sept. 1967, *Tweedie* 3487!
TANGANYIKA. Kigoma District: near Kigoma, Buhamba, 1 Feb. 1953, *Ross* 1509!; Iringa District: Sao Hill, 16 Aug. 1949, *Greenway* 8434!; Songea District: Kwamponjore valley, 9·5 km. SW. of Songea, 26 Apr. 1956, *Milne-Redhead & Taylor* 9830!
DISTR. U1, 4; K5; T4, 7, 8; Senegal to Zambia and Angola
HAB. Swamps, marshes and bogs; 900–2460 m.

SYN. *U. tortilis* Oliv. in J.L.S. 9: 150 (1865); Stapf in F.T.A. 4 (2): 483 (1906), pro parte

5. **U. baouleënsis** *A. Chev.* in Mém. Soc. Bot. Fr. 8: 186 (1912); P. Taylor in F.W.T.A., ed. 2, 2: 378 (1963) & in K.B. 18: 69 (1964) & in Fl. Afr. Centr., Lentib.: 6 (1972). Type: Ivory Coast, *Chevalier* 22247 (P, holo.!)

Terrestrial herb. Rhizoids and stolons capillary, few from the base of the scape. Leaves few, scattered on the stolons, linear, up to 3 cm. long, 0·4–1·0 mm. wide, 1-nerved. Traps few on the stolons and leaves, globose, 0·8–1·2 mm. long, shortly stalked, glandular; mouth basal; upper lip with 2 subulate branched appendages. Inflorescence twining, up to 20 cm. high; scape filiform, glabrous; flowers 2–5, distant; scales few, similar to the bracts; bracts basifixed, ovate, ± 1·2 mm. long; bracteoles linear-lanceolate, ± half as long as the bract; pedicels erect at anthesis, strongly recurved in fruit, ± as long as the calyx, flattened and narrowly winged. Calyx-lobes ovate, subequal, ± 2 mm. long at anthesis, 3·5–4 mm. long, with the lower lobe slightly relatively longer in fruit. Corolla pale blue or mauve, 3–4 mm. long; upper lip oblong, apex truncate, slightly longer than the upper calyx-lobe; lower lip orbicular, apex obscurely 3-crenate; palate scarcely raised; spur narrowly conical, obtuse. Filaments linear; anther-thecae subdistinct. Ovary ovoid; style short, distinct; stigma lower lip orbicular, upper lip very short, truncate. Capsule broadly ovoid, dehiscing by a single longitudinal abaxial slit with thickened margins in the upper half. Seeds numerous, ovoid to ellipsoid, ± 0·3 mm. long; testa loose, corky, the cells distinct, elongate.

Uganda. Teso District: Omunyal Swamp, 14 Sept. 1954, *Lind* 413!
Tanganyika. Uzaramo District: Fungoni, June 1965, *Procter* 3023!; Ulanga District: Ukindu, 25 June 1932, *Schlieben* 2523!
Distr. **U3**; **T6**; scattered throughout tropical Africa and in Madagascar, India (Madras), Philippines and Australia (Queensland)
Hab. Marshes and swampy grassland; 150–1000 m.

6. **U. pentadactyla** *P. Taylor* in Mem. N.Y. Bot. Gard. 9: 16 (1954) & in K.B. 18: 129 (1964) & in Fl. Afr. Centr., Lentib.: 14 (1972). Type: Rhodesia, *Wild* 3240 (K, holo.!, SRGH, iso.!)

Terrestrial herb. Rhizoids and stolons capillary, few from the base of the scape. Leaves often decayed at anthesis, or 2–5 subrosulate at the scape-base and scattered on the stolons, oblanceolate-spathulate, up to 6 mm. long and 1 mm. wide, 1-nerved. Traps numerous on the rhizoids and leaves, globose, 0·6–0·8 mm. long, shortly stalked; mouth terminal; upper lip projecting, lower lip not or scarcely projecting, both provided with radiating comb-like rows of glandular hairs. Inflorescence erect, 2–30 cm. high; scape filiform, glabrous above, papillose or setulose at the base; flowers 1–4, distant; scales few, similar to the bracts; bracts basifixed, ovate-deltoid, ± 0·6 mm. long; bracteoles similar but narrower; pedicels capillary, erect, shorter than the calyx-lobes. Calyx-lobes unequal; upper broadly ovate, 1–1·6 mm. long, obtuse to subacute; lower smaller, oblong, emarginate. Corolla pale mauve to white with a yellow spot on the palate, 3–15 mm. long; upper lip narrowly oblong, 1·5–2·5 times as long as the upper calyx-lobe, apex emarginate to deeply bifid; lower lip orbicular in outline, ± deeply 5-lobed; palate raised, often crested; spur subulate, slightly curved, up to 10 mm. long. Filaments filiform; anther-thecae subdistinct. Ovary ovoid; style very short; stigma lower lip semi-orbicular, upper lip much smaller, deltoid. Capsule globose, 1·5–2 mm. long, dehiscing by a longitudinal abaxial slit with thickened margins. Seeds numerous, somewhat conical, slightly angular; testa smooth, cells indistinct, elongated on the lateral surfaces, ± isodiametric elsewhere.

Uganda. Lango District: Orumo, Sept. 1935, *Eggeling* 2206!; Mbale District: Sipi, 3 Aug. 1932, *A. S. Thomas* 455! & Buginyanya, 1 Sept. 1932, *A. S. Thomas* 455A!
Kenya. Trans-Nzoia District: Kitale, July 1954, *Tweedie* 1789!
Tanganyika. Ufipa District: Mwimbi Dambo, 21 Apr. 1962, *Richards* 16352!; Rungwe District: Mbeye, *St. Clair-Thompson* 790!
Distr. **U1**, 3; **K3**; **T4**, 7; Ethiopia to Malawi and Rhodesia
Hab. Damp sandy or peaty grassland; 1500–2100 m.

7. **U. firmula** *Oliv.* in J.L.S. 9: 152 (1865); Stapf in F.T.A. 4 (2): 479 (1906); P. Taylor in F.W.T.A., ed. 2, 2: 378 (1963) & in K.B. 18: 151 (1964) & in Fl. Afr. Centr., Lentib.: 16, t. 3 (1972). Type: Angola, *Welwitsch* 262 (LISU, holo.!, BM, COI, G, K, P, iso.!)

Terrestrial herb. Rhizoids and stolons capillary, numerous from the base of the scape. Leaves scattered on the stolons, narrowly obovate-spathulate, 2–20 mm. long, ± 1 mm. wide, 1-nerved. Traps numerous, ovoid, 0·15–0·3 mm. long, stalked, the stalk ± as long as the trap; mouth terminal; upper lip projecting, fringed with 2–10 glandular hairs. Inflorescence stiffly erect, simple or sometimes branched above, 2–36 cm. high; scape relatively thick and rigid, glabrous; flowers (1–)10–20(–50); scales numerous, especially below, similar to the bracts; bracts basifixed, ovate-deltoid, reflexed;

bracteoles lanceolate, erect; pedicels erect, 0·2–0·5 mm. long. Calyx-lobes subequal, orbicular, very concave, apex emarginate or shortly 2–3-denticulate. Corolla pale yellow, 3·5–6 mm. long, persistent; upper lip broadly ovate, apex emarginate, ± 1·5 times as long as the upper calyx-lobe; lower lip longer, orbicular in outline, distinctly 3-lobed; palate slightly raised; spur subulate, 2·5–3·5 times as long as the lower lip. Filaments filiform, curved; anther-thecae ± confluent. Ovary ovoid; style very short; stigma lower lip rounded, upper lip minute, deltoid. Capsule globose, ± 1·2 mm. long, dehiscing by ab- and adaxial ovate-lanceolate pores. Seeds numerous, narrowly truncate-conical; testa smooth, cells indistinct, elongate.

UGANDA. Acholi District: Kilak Hill, 19 Nov. 1941, *A. S. Thomas* 4050!; Mbale District: Sipi, 31 Aug. 1932, *A. S. Thomas* 456!
KENYA. Trans-Nzoia District: Kitale, 18 Sept. 1956, *Bogdan* 4300!; Uasin Gishu District: Kipkarren, Mar. 1932, *Brodhurst Hill* 715!; Kwale District: Shimba Hills, 17 Mar. 1968, *Magogo & Glover* 321!
TANGANYIKA. Mpanda District: Kasimba, July 1951, *Eggeling* 6168!; Rufiji District: Mafia I., 7 Aug. 1936, *FitzGerald* 5218!; Songea District: 6·5 km. W. of Songea, 3 May 1956, *Milne-Redhead & Taylor* 9882!
ZANZIBAR. Zanzibar I., Kama Swamp, 4 Sept. 1963, *Faulkner* 3268!
DISTR. U1, 3; K3, 7; T4–8; Z; P; throughout tropical and subtropical Africa and in Madagascar
HAB. Boggy grassland; sea-level to 1860 m.

8. **U. welwitschii** *Oliv.* in J.L.S. 9: 152 (1865); Stapf in F.T.A. 4 (2): 478 (1906); P. Taylor in K.B. 18: 144 (1964), pro parte, excl. vars. & in Fl. Afr. Centr., Lentib.: 18, t. 4/1–18 (1972). Type: Angola, *Welwitsch* 266 (LISU, holo.!, BM, C, COI, G, K, P, iso.!)

Terrestrial herb. Rhizoids and stolons capillary, numerous from the base of the scape. Leaves scattered on the stolons, often decayed at anthesis, obovate-spathulate to linear-oblanceolate, up to 3 cm. long, 0·5–2 mm. wide, 1-nerved. Traps numerous, ovoid, 0·4–0·8 mm. long, stalked; mouth terminal; upper lip projecting, fringed with ± 5 glandular hairs; lower lip with radiating rows of ± sessile glands. Inflorescence erect, simple or branched above, 4–50 cm. high; scape usually robust, straight or twining, smooth; inflorescence-axis usually flexuous; flowers 1–25 or more, usually distant; scales numerous, similar to the bracts; bracts basifixed, lanceolate to ovate, 0·6–1·6 mm. long; bracteoles lanceolate, ± as long as the bracts; pedicels up to 1 mm. long. Calyx-lobes subequal, orbicular, 1·5–2·5 mm. long, densely papillose, margin strongly recurved, apex tridenticulate. Corolla usually violet with a yellow blotch on the palate, sometimes wholly yellow, 5–15 mm. long; upper lip broadly ovate to orbicular, 2–4 times as long as the upper calyx-lobe, apex rounded, truncate or emarginate; lower lip orbicular, apex rounded, entire or obscurely 3-crenate; palate raised, usually distinctly 2-gibbous; spur subulate, acute, 1·5–2 times as long as the lower lip. Filaments filiform; anther-thecae subdistinct. Ovary globose; style obsolete; stigma lower lip semi-orbicular, upper much smaller, deltoid. Capsule globose, ± 1·5 mm. long, dehiscing by ab- and adaxial ovate-lanceolate pores. Seeds numerous, ovoid, angular, 0·25 mm. long; testa smooth, cells indistinct, ± isodiametric.

TANGANYIKA. Kigoma District: Kafulu, July 1951, *Eggeling* 6178!; Iringa District: Dabaga Highlands, 21 Feb. 1962, *Polhill & Paulo* 1556!; Songea District: Luhira valley, 6 Nov. 1956, *Semsei* 2615a!
DISTR. T4, 7, 8; Zaire (Katanga), Rwanda and Burundi to Angola and South Africa (Transvaal), also in Madagascar
HAB. Marshy grassland; 900–2000 m.

SYN. *U. welwitschii* Oliv. var. *welwitschii*; P. Taylor in K.B. 18: 146 (1964)

9. **U. arenaria** *A. DC.* in DC., Prodr. 8: 20 (1844); P. Taylor in F.W.T.A., ed. 2, 2: 378 (1963) & in K.B. 18: 107 (1964) & in Fl. Afr. Centr., Lentib.: 22 (1972). Type: Senegal, *Perrottet* (G, holo.!)

Terrestrial herb. Rhizoids and stolons capillary, numerous from the base of the scape. Leaves usually numerous at anthesis, scattered on the stolons, linear-oblanceolate to obovate-spathulate, 2–15 mm. long, 0·3–2 mm. wide, 1-nerved. Traps numerous, ovoid, stalked, 0·6–1·0 mm. long; mouth terminal; upper lip projecting ± twice as much as the lower, both provided with radiating rows of glandular hairs. Inflorescence erect, 2–16 cm. high; flowers 1–5(–8), distant; scape filiform, smooth above, usually papillose at the base; scales few, similar to the bracts; bracts basifixed, ovate-lanceolate, acute, ± 1 mm. long; bracteoles similar but narrower; pedicels 0·5–1 mm. long. Calyx-lobes subequal, the upper slightly larger, broadly ovate, acute, lower ovate-oblong, rounded or truncate. Corolla white or lilac with a yellow spot on the palate, 3–7 mm. long; upper lip narrowly oblong, ± 1·5 times as long as the upper calyx-lobe, apex rounded, truncate or emarginate; lower lip orbicular; palate raised, double-crested, the crests smooth or transversely tuberculate; spur conical-subulate, ± twice as long as the lower lip. Filaments filiform; anther-thecae subdistinct. Ovary ovoid; style very short; stigma lower lip semi-orbicular, upper much smaller, deltoid. Capsule globose, 1·5–2·5 mm. in diameter, dehiscing by longitudinal ab- and adaxial slits with thickened margins. Seeds numerous, ± 0·2 mm. long, truncate-conical, angular; testa smooth, cells indistinct, oblong.

UGANDA. Acholi District: Gulu, 13 Nov. 1941, *A. S. Thomas* 4016!; Mbale District: Sipi, 31 Aug. 1932, *A. S. Thomas* 454!; Masaka District: Lake Nabugabo, Aug. 1935, *Chandler* 1283!
KENYA. Trans-Nzoia District: Kitale, 6 Oct. 1958, *Bogdan* 4666!; N. Kavirondo District: Broderick Falls, Aug. 1964, *Tweedie* 3083!; Kilifi District: Mida, 3 Dec. 1961, *Polhill & Paulo* 902!
TANGANYIKA. Moshi District: between Moshi and Arusha, 15 Dec. 1961, *Polhill & Paulo* 997!; Tanga District: Nyamaku, 21 July 1957, *Faulkner* 2031!; Songea District: R. Luhira N. of Songea, 18 Mar. 1956, *Milne-Redhead & Taylor* 9237!
ZANZIBAR. Zanzibar I., Kama Swamp, 4 Sept. 1963, *Faulkner* 3269!
DISTR. **U**1, 3, 4; **K**3–5, 7; **T**1–3, 6–8; **Z**; **P**; Senegal to Ethiopia and south to South West Africa and South Africa (Natal), Madagascar and India
HAB. Boggy grassland, damp open sandy ground and rock pavements; sea-level to 2650 m.

SYN. *U. tribracteata* A. Rich., Tent. Fl. Abyss. 2: 18 (1851); Stapf in F.T.A. 4 (2): 475 (1906), pro parte. Type: Ethiopia, *Schimper* 1943 (FI, G, K, M, MO, P, S, W iso.!)
    U. exilis Oliv. in J.L.S. 9: 154 (1865); Stapf in F.T.A. 4 (2): 477 (1906). Types: Angola, *Welwitsch* 252, 253, 254 & 256 (LISU, syn.!, BM, C, COI, G, K, P, Z, isosyn.!)
    U. kirkii Stapf in Fl. Cap. 4 (2): 428 (1904) & in F.T.A. 4 (2): 476 (1906), pro parte. Types: South Africa, Transvaal, *Rehmann* 5699 & s.n. (Z, isosyn.!)

10. **U. livida** *E. Mey.*, Comm. Pl. Afr. Austr.: 281 (1837); P. Taylor in K.B. 18: 115 (1964) & in Fl. Afr. Centr., Lentib.: 24 (1972). Type: South Africa, Cape Province, *Drège* 4838 (K, iso.!)

Terrestrial herb. Rhizoids and stolons capillary, usually numerous from the base of the scape. Leaves not always present nor conspicuous at anthesis, subrosulate at the scape-base and scattered on the stolons, linear to obovate-spathulate or rarely reniform, 1–7 cm. long, 1–6 mm. wide; nerves of the lamina usually dichotomously branched. Traps numerous, ovoid, 1–2 mm. long, stalked; mouth terminal; upper lip projecting about twice as much as the lower, both provided with radiating rows of glandular hairs. Inflorescence erect, straight or flexuous, simple or rarely branched above, 2–80 cm. high;

scape smooth and glabrous, relatively robust; flowers (1–)2–8(–50), distant or more rarely congested; scales few, similar to the bracts; bracts basifixed, ovate, acute or acuminate, ± 1 mm. long; bracteoles linear-lanceolate, ± as long as the bract; pedicels 0·5–1(–3) mm. long, erect at anthesis, spreading or reflexed in fruit. Calyx-lobes subequal, ovate, 2–3 mm. long at anthesis, accrescent, always plicate along the nerves, apex of upper lobe acute or obtuse, of lower lobe rounded, truncate or ± bidentate. Corolla violet, mauve or white with a yellow spot on the palate or more rarely wholly yellow or cream, 5–15 mm. long; upper lip 1·5–2 times as long as the upper calyx-lobe, narrowly oblong-obovate, apex rounded or truncate; lower lip orbicular; palate raised and double crested, the crests usually transversely tuberculate; spur slightly shorter than to about 1½ times as long as the lower lip, conical-subulate, straight or curved. Filaments linear; anther-thecae subdistinct. Ovary globose; style short but distinct; stigma lower lip semi-orbicular, upper much smaller, deltoid. Capsule globose, up to 2 mm. long, dehiscing by ab- and adaxial longitudinal slits with thickened margins. Seeds few to many, 0·3–0·5 mm. long, ovoid, slightly angular, smooth or obscurely to distinctly papillose; testa-cells indistinct, elongate.

UGANDA. Karamoja District: Mt. Kadam, Nov. 1964, *J. Wilson* 1621 !; Kigezi District: Buhara, 2 Sept. 1952, *Norman* 146 !
KENYA. Trans-Nzoia District: Kitale, 5 Oct. 1958, *Bogdan* 4663 !; Mt. Kenya, Nithi R., 11 Jan. 1958, *Coe & Kirrika* 382 !; Masai District: Ol Donyo Orok, 31 May 1971, *Archer* 666 !
TANGANYIKA. Uluguru Mts., 25 Sept. 1932, *Schlieben* 2716 !; Iringa District: Dabaga Highlands, 9 Feb. 1962, *Polhill & Paulo* 1412 !; Tunduru District: W. of Puchapucha, 19 Dec. 1955, *Milne-Redhead & Taylor* 7715 !
DISTR. **U**1, 2; **K**1–4, 6; **T**2–8; Ethiopia to Cape Province, Madagascar and Mexico
HAB. Boggy grassland and wet rocky places; sea-level to 2600 m.

SYN.   *U. transrugosa* Stapf in Fl. Cap. 4 (2): 428 (1904) & in F.T.A. 4 (2): 473 (1906). Types: South Africa, Transvaal, *Galpin* 520 (K, syn. !, BOL, isosyn. !) & *Rand* 727 (BM, syn. !) & *Ommaney* 129 (BM, syn. !, MO, isosyn. !) & *Pegler* 936 (K, syn. !, BOL, PRE, Z, isosyn. !)
      U. sematophora Stapf in E.J. 40: 60 (1907). Type: Tanganyika, Njombe District, Ukinga Mts., *Goetze* 976 (BR, K, iso. !)
      U. afromontana R. E. Fries in N.B.G.B. 8: 703 (1924). Type: Mt. Kenya, *Fries* 407 (BR, K, UPS, S, iso. !)

11. **U. pubescens** *Sm.* in Rees, Cycl. 37, No. 53 (1819); P. Taylor in F.W.T.A., ed. 2, 2 : 378 (1963) & in K.B. 18 : 101 (1964) & in Fl. Afr. Centr., Lentib.: 29 (1972). Type: Sierra Leone, *Afzelius* (LINN, holo. !, BM, BR, S, iso. !)

Terrestrial herb. Rhizoids and stolons capillary, numerous from the base of the scape. Leaves usually present at anthesis, scattered on the stolons, orbicular, peltate, long-petiolate; lamina fleshy, mucilaginous, 1–5 mm. in diameter; petiole 2–10 mm. long. Traps numerous, globose, stalked, 0·5–0·8 mm. long; mouth terminal; upper lip projecting; lower lip scarcely projecting, both provided with radiating rows of glandular hairs. Inflorescence erect, 2–35 cm. high; scape filiform, straight, ± setulose, papillose or rarely entirely glabrous; flowers 1–6(–10), distant; scales few, similar to the bracts; bracts basisolute, shortly produced below the point of insertion, ovate, apex acute, base truncate, ± 1 mm. long, setulose; bracteoles similar to the bracts but narrower; pedicels erect, 0·5–2 mm. long. Calyx-lobes subequal, broadly ovate, usually setulose; upper lobe ovate, 1·5–3 mm. long, apex acute; lower lobe ovate or orbicular, apex bidentate. Corolla white or pale lilac, 2–16 mm. long; upper lip 1·5–2 times as long as the upper calyx-lobe, narrowly oblong, apex rounded, truncate or ± emarginate; lower lip orbicular, entire; palate much raised, double-crested, the crests usually transversely tuberculate; spur usually conical-subulate, up to 4 times as long as the lower lip but very variable in length and sometimes much reduced. Filaments filiform; anther-thecae subdistinct. Ovary ovoid; style short but usually distinct; stigma lower lip semi-orbicular, upper much smaller, deltoid. Capsule globose dehiscing by a longitudinal abaxial slit with thickened margins. Seeds numerous, ovoid; testa-cells distinct, ± isodiametric.

UGANDA. Masaka District: Lake Nabugabo, 7 Oct. 1953, *Drummond & Hemsley* 4676!
TANGANYIKA. Bukoba, Jan. 1932, *Haarer* 2439!; Kigoma District: Buhamba, 1–2 Feb. 1953, *Ross* 1508! & 1516!; Songea District: 12 km. W. of Songea, 30 Dec. 1955, *Milne-Redhead & Taylor* 7795!
DISTR. U4; T1, 4, 8; Guinée to Zambia and Angola, also in S. America and India
HAB. Boggy grassland and wet rocks; 780–1200 m.

SYN. *U. fernaldiana* Lloyd & G. Taylor in Contr. Gray Herb. 165: 87 (1947). Type: Uganda, Masaka District, Lake Nabugabo, *A. S. Thomas* 1335 (BM, holo.!)
    U. thomasii Lloyd & G. Taylor in Contr. Gray Herb. 165: 88 (1947). Type: Uganda, Masaka District, Lake Nabugabo, *A. S. Thomas* 1334 (BM, holo.!)

12. **U. striatula** *Sm.* in Rees, Cycl. 37, No. 17 (1819); Stapf in F.T.A. 4 (2): 486 (1906); P. Taylor in F.W.T.A., ed. 2, 2: 378 (1963) & in K.B. 18: 91 (1964) & in Fl. Afr. Centr., Lentib.: 32, t. 9 (1972). Type: Sierra Leone, *Afzelius* (LINN, holo.!, BM, BR, LD, S, iso.!)

Epiphytic herb. Rhizoids and stolons capillary, numerous from base of the scape. Leaves numerous and conspicuous at anthesis, rosulate at the scape-base and scattered on the stolons, obovate to reniform, petiolate, 3–10 mm. total length, the lamina 1–6 mm. wide. Traps numerous, globose or ovoid, 0·6–0·8 mm. long, stalked; mouth lateral; upper lip shortly projecting with 2 divergent appendages densely covered with glandular hairs. Inflorescence erect, straight, 1–15 cm. high; scape capillary, glabrous; flowers 1–10, distant; scales few, similar to the bracts; bracts medifixed, lanceolate, 1·5–2 mm. long; bracteoles similar to the bracts but smaller; pedicels capillary, 2–6 mm. long, spreading or sometimes reflexed in fruit. Calyx-lobes very unequal and accrescent; upper lobe orbicular-obcordate, apex emarginate, 1·5–2·5 mm. long at anthesis; lower lobe ovate-oblong, ± half as long as the upper. Corolla white or mauve with a yellow spot on the throat, 3–6 mm. long or sometimes much reduced (cleistogamous); upper lip minute, much shorter than the upper calyx-lobe, very shortly 2-lobed; lower lip orbicular in outline, 3–10 mm. long, ± regularly 5-lobed; palate slightly raised; spur subulate, curved or straight, ± as long as the lower lip. Filaments filiform; anther-thecae subdistinct. Ovary globose, adnate to the base of the upper calyx-lobe; style short; stigma lower lip semi-orbicular, upper lip obsolete. Capsule globose, obliquely dorsiventrally compressed, carinate on the abaxial surface, dehiscing by a longitudinal abaxial slit. Seeds numerous, ovoid, 0·25 mm. long, covered with short glochidiate processes.

UGANDA. Mbale District: Buginyanya, 2 Sept. 1932, *A. S. Thomas* 469! & Kapchorwa, 9 Sept. 1954, *Lind* 475!
TANGANYIKA. Mpanda District: Mahali Mts., 15–17 July 1959, *Newbould & Harley* 4359! & 4389!; Morogoro District: N. Uluguru Mts., Mambi falls near Mzinga, 7 Jan. 1970, *T. & S. Pócs* 6110/D!; Iringa District: Gologolo Mts., 13 Sept. 1970, *Thulin* 959!
DISTR. U3; T4, 6, 7; Guinée to Zambia and Angola, but apparently absent from Zaire and Madagascar, widespread in tropical Asia to New Guinea
HAB. Wet rocks and tree-trunks among mosses; 1500–2250 m.

13. **U. appendiculata** *E. A. Bruce* in K.B. 1933: 475 (1934); P. Taylor in K.B. 18: 95 (1964) & in Fl. Afr. Centr., Lentib.: 30, t. 8 (1972). Type: Tanganyika, Biharamulo, *Haarer* 2056 (K, holo.!, EA, iso.!)

Terrestrial herb. Rhizoids and stolons capillary, numerous from the base of the scape. Leaves rosulate but usually decayed at anthesis, linear, thalloid, up to 3 cm. long, 1·5–3 mm. wide, 1-nerved. Traps numerous, ovoid, 0·8–1·0 mm. long, stalked; mouth lateral, oblique; upper lip with a single variable but usually 3-branched appendage ± recurved over the mouth. Inflorescence filiform, twining, up to 60 cm. high, glabrous above, minutely papillose at the base; flowers 1–8, distant; scales numerous, similar to the bracts; bracts medifixed, ± 3 mm. long, narrowly lanceolate, acute at both extremities; bracteoles similar but somewhat smaller; pedicels erect, 2–3 mm. long. Calyx-lobes very unequal; upper broadly rhomboid, ± 3 mm. long; lower ovate or ovate-oblong, much narrower than the upper, apex truncate. Corolla white, cream or pale yellow, 5–9 mm. long; upper lip oblong, scarcely exceeding the upper calyx-lobe, apex truncate or emarginate; lower lip orbicular, apex entire or emarginate; palate raised, obscurely 2-gibbous; spur narrowly cylindrical, curved, ± as long as the lower lip. Filaments filiform, twisted; anther-thecae subdistinct. Ovary ovoid; style very short; stigma lower lip semi-orbicular, upper much smaller, deltoid or oblong. Capsule globose, obliquely dorsiventrally compressed, dehiscing by a longitudinal abaxial slit. Seeds very numerous, narrowly cylindrical; testa longitudinally striate and with numerous short rough conical papillae.

UGANDA. Kigezi District: Kashambya Swamp, 6 Sept. 1952, *Norman* 155!; Masaka District: Lake Nabugabo, 6 Oct. 1953, *Drummond & Hemsley* 4655!
TANGANYIKA. Biharamulo, Aug. 1931, *Haarer* 2056!; Iringa District: Sao Hill, Apr. 1959, *Watermeyer* 91!; Songea District: Matengo Hills, 1·5 km. N. of Miyau, 2 Mar. 1956, *Milne-Redhead & Taylor* 8799!
DISTR. U2, 4; T1, 7, 8; Cameroun to Mozambique and Madagascar
HAB. *Sphagnum* bogs and boggy grassland; 1140–1860 m.

14. **U. subulata** *L.*, Sp. Pl.: 18 (1753); P. Taylor in F.W.T.A., ed. 2, 2: 380 (1963) & in K.B. 18: 81 (1964) & in Fl. Afr. Centr., Lentib.: 34 (1972). Type: U.S.A., *Clayton* (BM, holo.!)

Terrestrial herb. Rhizoids and stolons capillary, few from the base of the scape. Leaves linear, 1–2 cm. long, ± 0·5 mm. wide, 1-nerved. Traps very numerous, globose or ovoid, 0·2–0·5 mm. long, stalked, the stalk slightly shorter than the trap; mouth lateral, oblique; upper lip with 2 branched appendages. Inflorescence erect, up to 25 cm. high; scape capillary, smooth above, minutely papillose below; inflorescence-axis usually flexuous or zigzag; floral internodes slightly longer than the pedicels; scales few, peltate, ± as long as the bracts, narrowly elliptic, acuminate at both extremities, minutely papillose; bracts peltate, orbicular, 0·75–1·0 mm. long, smooth, clasping the base of the pedicel; bracteoles absent; pedicels capillary, ascending, 2–10 mm. long. Calyx-lobes subequal, broadly ovate, ± 1 mm. long at anthesis, slightly accrescent, obscurely 5-nerved, apex obtuse or truncate. Corolla yellow, normally 6–10 mm. long, sometimes reduced (in cleistogamous flowers) to 2 mm.; upper lip broadly ovate, apex rounded, 2–3 times as long as the upper calyx-lobe; lower lip orbicular in outline, deeply 3-lobed; palate raised, bigibbous; spur subulate, parallel with and ± equalling the lower lip or (in cleistogamous flowers) much reduced and saccate. Filaments filiform; anther-thecae confluent. Ovary globose; style very short; stigma lower lip orbicular, upper lip ± obsolete. Capsule globose, 1–1·5 mm. long, dehiscing by a small ovate abaxial pore. Seeds numerous, ovoid, 0·2–0·24 mm. long, longitudinally striate.

UGANDA. W. Nile District: Koboko, *Eggeling* 1855!; Mbale District: Sipi, 31 Aug. 1932, *A. S. Thomas* 457!; Masaka District: Lake Nabugabo, 6 Oct. 1953, *Drummond & Hemsley* 4640!

Tanganyika. Bukoba, June 1931, *Haarer* 2016!; Kigoma District: Buhamba, 1 Feb. 1953, *Ross* 1507!; Rufiji District: Mafia I., 6 Aug. 1936, *FitzGerald* 5212/1!; Songea District: Nangurukuru Hill, 8 Apr. 1956, *Milne-Redhead & Taylor* 9490!

Distr. U1, 3, 4; T1, 4, 6, 8; throughout tropical Africa to South West Africa and South Africa (Transvaal, Natal), Madagascar, Portugal, America from Nova Scotia to Argentina and in Thailand and Borneo

Hab. Wet peaty or sandy soil; sea-level to 2250 m.

15. U. benjaminiana *Oliv.* in J.L.S. 4: 176 (1860); P. Taylor in F.W.T.A., ed. 2, 2: 380 (1963) & in K.B. 18: 179 (1964) & in Fl. Afr. Centr., Lentib.: 36, t. 10 (1972). Type: S. America, Suriname, *Hostmann* 85 (K, holo.!, BM, NY, P, iso.!)

Aquatic herb. Stolons filiform, up to 50 cm. long or more, flexuous, ± densely villous; internodes 1–10 cm.; rhizoids absent. Leaves with 1 or 2 primary segments 1·5–4 cm. long, ovate in outline, pinnately divided; rhachis filiform throughout or ± inflated, especially in the lower half, villous; pinnae alternate, the lowermost up to 1·5 mm. from the base of the rhachis, repeatedly dichotomously forked, ultimate segments capillary, sparsely setulose. Floats terminal on short lateral slightly inflated villous branches, verticillate, 4(–5–6)–9, fusiform or narrowly cylindrical to ovoid or ellipsoid, 0·3–2·5 cm. long, bearing in the distal half a number of leaf-segments. Traps usually not numerous and often absent from some leaves, inserted in the angles between the leaf-rhachis and pinnae and between subsequent dichotomies of the leaves, ovoid, long-stalked, villous, 1–3 mm. long; mouth lateral; upper lip with 2 simple or ± branched hairs; lower lip naked or with 1 or more shorter simple hairs. Inflorescence erect, 3–25 cm. high, terminal on the short stolon-branches, arising from the centre of the whorl of floats; flowers 2–10, the uppermost congested, those below ± distant; cleistogamous flowers often present at the scape-base among the floats; scape filiform, straight or flexuous, smooth and glabrous; scales absent; bracts basifixed, deltoid or ovate, ± 1·5 mm. long; bracteoles absent; pedicels capillary, erect, 2–3 mm. long at anthesis, up to 10 mm. long in fruit. Calyx-lobes subequal, orbicular, membranous, ± 1 mm. long at anthesis, scarcely accrescent. Corolla mauve or pale purple with a yellow blotch on the palate or wholly white, 10–15 mm. long; upper lip oblong, 3–4 times as long as the upper calyx-lobe, ± auriculate at the base, deeply divided into 2 parallel lobes; lower lip reniform, 7–10 mm. wide, apex ± emarginate; palate with a narrow raised ± crenulate rim; spur cylindrical or botuliform, 2–3 times as long as the lower lip, 8–10 mm. long, 2·5–3·5 mm. thick; corolla of cleistogamous flowers obsolete or nearly so. Filaments filiform; anther-thecae confluent. Ovary ovoid; ovules 3–20; style short but distinct; stigma lower lip ± quadrate, upper minute, deltoid or ± obsolete. Capsule ellipsoid, 2·5–3·5 mm. long, circumscissile. Seeds few, sometimes only 2, lenticular with a narrow irregular wing, ± 1 mm. in total diameter; testa-cells indistinct, elongated.

Uganda. Masaka District: Lake Nabugabo, 5 Mar. 1933, *A. S. Thomas* 959!

Tanganyika. Bukoba District: Nyakato, Aug. 1931, *Haarer* 2120!; Rufiji District: Mafia I., 7 Aug. 1937, *Greenway* 5021!

Zanzibar. Pemba I., Matanga Twani [Mtangatwani], 27 Sept. 1929, *Vaughan* 679!

Distr. U4; T1, 6; Z; P; throughout tropical Africa from Senegal to South West Africa and South Africa (Natal), Madagascar and E. tropical S. America

Hab. Still and slow flowing water in pools, marshes and rivers; sea-level to 1200 m.

Syn. *U. gilletii* De Wild. & Th. Dur. in B.S.B.B. 38, Compt. Rend.: 40 (1900). Type: Zaire, *Gillet* (BR, holo.!)
    U. villosula Stapf in F.T.A. 4 (2): 490 (1906); F.W.T.A. 2: 234 (1931). Type: Angola, R. Longa, *Baum* 656 (K, holo.!, BM, COI, G, HBG, M, S, W, Z, iso.!)

16. **U. inflexa** *Forsk.*, Fl. Aegypt.-Arab., Descr. Pl.: 9 (1775); P. Taylor in
Fl. Afr. Centr., Lentib.: 38, t. 11/1–5 & t. 12 (1972). Type: Egypt, Rosetta,
Forsskål (C, holo. !, P, iso. !)

Aquatic herb. Stolons up to 1 m. long or more, terete, smooth and glabrous;
internodes 3–10(–15) mm. Leaves very numerous, digitately divided into 3–6
primary segments and usually auricled at the base; primary segments
2–6 cm. long, ± lanceolate in outline, pinnately divided, the rhachis filiform
or ± inflated and up to 2 mm. thick; pinnae repeatedly dichotomously
forked; ultimate segments capillary, minutely setulose; auricles when
present reniform or orbicular, up to 10 mm. long, margin usually denticulate
or ± divided into setulose capillary segments. Floats verticillate or subverticillate on the lower half of the scape, 3–10, inflated, narrowly cylindrical
to fusiform, 2–5 cm. long, bearing a variable number of reduced or sometimes well-developed leaf-segments at the apex. Traps usually numerous,
lateral near the base of the ultimate and penultimate leaf-segments, broadly
ovoid, shortly stalked, 1–3 mm. long; mouth lateral; upper lip naked or
with 2 short simple or sparsely branched hairs; lower lip naked or with a
few shorter simple hairs. Inflorescence erect, lateral, 3–20 cm. high; flowers
2–15, congested at anthesis; inflorescence-axis elongating in fruit; scape
filiform to relatively stout, straight, smooth and glabrous; scales absent;
bracts basifixed, ovate, 1–2 mm. long; bracteoles absent; pedicels filiform,
erect or spreading at anthesis, 1·5–5 mm. long, strongly reflexed and thickened
in fruit. Calyx-lobes subequal, broadly ovate, apex rounded or lower sometimes emarginate, ± 3 mm. long at anthesis, strongly accrescent, becoming
fleshy, up to 10 mm. long and enclosing the capsule. Corolla white or yellow,
usually variously marked with red or purple lines, 7–10 mm. long, ± densely
covered externally with glandular hairs; upper lip broadly ovate, 1·5–2
times as long as the upper calyx-lobe, apex truncate or emarginate; lower lip
orbicular, apex rounded or emarginate; palate raised, bigibbous; spur
cylindrical, ± as long as the lower lip; filaments linear, dilated above;
anther-thecae subdistinct. Ovary globose, minutely squamulate; style
distinct, as long as the ovary; stigma lower lip semi-orbicular, upper obsolete.
Capsule globose, circumscissile, ± 5 mm. in diameter with the style often
considerably elongated. Seeds numerous, prismatic, 4–6-angled, ± as wide
as high, ± winged on the angles, hilum prominent; testa-cells distinct,
elongated, ± 0.05 mm. long.

UGANDA. W. Nile District: Laropi, *Eggeling* 915 !; Teso District: Pingire, 20 Sept. 1954,
    Lind 412 !; Masaka District: Sembabule, 23 July 1946, *A. S. Thomas* 4506 !
KENYA. Central Kavirondo District: NW. of Kisumu, 16 Sept. 1951, *Norman* 50 !;
    Masai District: Lake Natron, 29 July 1956, *Milne-Redhead & Taylor* 11311 !; Kwale
    District: Mwasangombe Forest, 27 Aug. 1953, *Drummond & Hemsley* 4017 !
TANGANYIKA. Arusha District: Ngongongare, 10 Dec. 1968, *Richards* 23315 !; Mpwapwa
    District: between Mtera and Chipogolo [Kipogoro], 19 Apr. 1962, *Polhill & Paulo*
    2082 !; Lindi District: Lake Lutamba, 26 Mar. 1935, *Schlieben* 6187 !
ZANZIBAR. Zanzibar I., Marahubi, 20 Sept. 1961, *Faulkner* 2906 !
DISTR. U1–4; K3–7; T1–6, 8; Z; throughout tropical Africa, extending to Madagascar
    and India
HAB. Still or slow flowing water in pools, lakes and rivers; sea-level to 1600 m.

SYN. *U. thonningii* Schumach. in Schumach. & Thonn., Beskr. Guin. Pl.: 12 (1827);
        Stapf in F.T.A. 4 (2): 487 (1906); F.W.T.A. 2: 234 (1931). Type: Ghana,
        Thonning (C, holo. !, W, iso. !)
    U. thonningii Schumach. var. *laciniata* Stapf in F.T.A. 4 (2): 488 (1906). Type:
        Zanzibar or the mainland opposite, *Kirk* 5 (K, holo. !)
    U. inflexa Forsk. var. *inflexa*; P. Taylor in K.B. 18: 186 (1964)

17. **U. stellaris** *Linn. f.*, Suppl.: 86 (1781); Stapf in F.T.A. 4 (2): 489
(1906); F.W.T.A. 2: 234 (1931); P. Taylor in Fl. Afr. Centr., Lentib.: 42,
t. 13 (1972). Type: India, *Koenig* (BM, holo. !, LD, iso. !)

Aquatic herb. Stolons up to 1 m. long or more and up to 2·5 mm. thick, terete, smooth and glabrous; internodes 5–20 mm. Leaves very numerous, digitately divided into 3–6 primary segments and usually auricled at the base; primary segments 1–6 cm. long, ± lanceolate in outline, pinnately divided; rhachis filiform or ± inflated, up to 2 mm. thick; pinnae repeatedly dichotomously forked; ultimate segments capillary, minutely setulose; auricles ± deeply divided into ciliate filiform segments. Floats verticillate or subverticillate on the upper half of the scape, 5–7, inflated, ellipsoid, 0·5–2 cm. long, bearing a few reduced leaf-segments at the apex. Traps usually numerous, lateral near the base of the ultimate and penultimate leaf-segments, broadly ovoid, shortly stalked, 1–3 mm. long; mouth lateral; upper lip naked or with 2 short simple or sparsely branched hairs; lower lip naked or with a few short simple hairs. Inflorescence erect, lateral, 3–30 cm. high; flowers 2–16, ± congested at anthesis; inflorescence-axis elongating with maturity; scape filiform, straight, smooth and glabrous; scales absent; bracts basifixed, broadly ovate, 2–3 mm. long; bracteoles absent; pedicels filiform, erect or spreading at anthesis, 1·5–5 mm. long, reflexed or recurved and ± elongating and thickening in fruit. Calyx-lobes subequal, broadly ovate, decurrent, ± 3 mm. long at anthesis, up to 5 mm. and reflexed in fruit; upper lobe apex rounded; lower lobe apex truncate or emarginate. Corolla yellow, usually marked with reddish lines on the palate, 7–10 mm. long, ± densely covered externally with glandular hairs; upper lip broadly ovate, 1·5–2 times as long as the upper calyx-lobe, apex truncate or emarginate; lower lip orbicular or oblate, truncate, emarginate or 3-crenate; palate raised, bigibbous; spur cylindrical, ± as long as the lower lip. Filaments linear, dilated above; anther-thecae subdistinct. Ovary globose; style short but distinct; stigma lower lip semi-orbicular, ciliolate, upper very short or obsolete. Capsule globose, circumscissile, ± 5 mm. in diameter with the style ± elongated. Seeds numerous, prismatic, 4–6-angled, ± 3 times as wide as high, usually narrowly winged on the angles; testa-cells distinct, elongated, ± 0·1 mm. long.

UGANDA. Acholi District: Chua, *Eggeling* 2416 !; Bunyoro District: Lake Albert flats near Sonso R., Dec. 1934, *Eggeling* 1483 !; Masaka District: Sembabule, 23 July 1946, *A. S. Thomas* 4505 !

KENYA. Kiambu District: Kabete, 30 Mar. 1962, *Kirrika* 447 !; Kisumu, *McMahon* 45 !; Tana River District: Tana R., Bura, 28 Sept. 1957, *Greenway* 9242 !

TANGANYIKA. Musoma District: Kleins Camp to Seronera, 6 Apr. 1961, *Greenway & Turner* 10004 !; Arusha District: Momela Lakes, 6 Apr. 1965, *Richards* 20062 !; Rufiji District: Mafia I., 2 Sept. 1937, *Greenway* 5208 !

DISTR. U1, 2, 4; K4, 5, 7; T1, 2, 4–7; throughout tropical and South Africa, Madagascar and tropical Asia to Australia

HAB. Still or slowly flowing water in lakes, marshes and rivers; sea-level to 1700 m.

SYN. *U. stellaris* Linn. f. var. *dilatata* Kam. in Ber. Deutsch. Bot. Ges. 12: 3 (1894). Types: Tanganyika, Tabora District, Wala R., *Boehm* 95 & 104 (B, syn. †)
U. trichoschiza Stapf in F.T.A. 4 (2): 488 (1906). Types: Nigeria, *Millsom & Mann* 2326 (both K, syn. !)
U. inflexa Forsk. var. *stellaris* (Linn. f.) P. Taylor in Mitt. Bot. Staats. München 4: 96 (1961) & in F.W.T.A., ed. 2, 2: 380 (1963) & in K.B. 18: 189 (1964)

18. **U. reflexa** *Oliv.* in J.L.S. 9: 146 (1865); Stapf in F.T.A. 4 (2): 492 (1906); P. Taylor in F.W.T.A., ed. 2, 2: 380 (1963) & in K.B. 18: 163 (1964) & in Fl. Afr. Centr., Lentib.: 45 (1972). Type: Angola, Huila, *Welwitsch* 269 (LISU, holo. !, BM, COI, G, K, P, iso. !)

Aquatic herb. Stolons filiform to relatively thick and fleshy, up to 50 cm. long or more, 0·3–3·0 mm. thick, glabrous or ± densely covered with simple or stellate hairs which extend also to the traps, basal part of the leaves and scapes; internodes 2–10(–15) mm. Rhizoids usually absent but when present

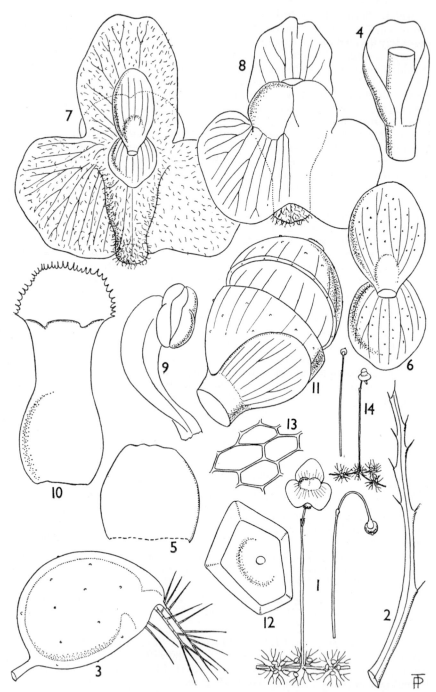

FIG. 2. *UTRICULARIA REFLEXA* var. *REFLEXA*—**1**, part of flowering plant with detached fruiting scape, × 1; **2**, ultimate leaf-segment, × 30; **3**, trap, × 30; **4**, bract, in situ, × 12; **5**, bract, flattened, × 12; **6**, calyx, × 12; **7**, flower, adaxial view, × 8; **8**, corolla, abaxial view, × 8; **9**, stamen, × 30; **10**, pistil, × 30; **11**, dehisced capsule, × 12; **12**, seed, × 30; **13**, testa cells, × 150. *U. REFLEXA* var. *PARVI-FLORA*—**14**, part of flowering plant with detached fruiting scape, × 1. 1, from *Milne-Redhead & Taylor* 9831; 2, 11, from *Milne-Redhead & Taylor* 9832; 3–7, 9, 10, from *Milne-Redhead & Taylor* 8066a; 8, from *Milne-Redhead & Taylor* 11121; 12, 13, from *Lind* 415; 14, from *Milne-Redhead & Taylor* 9236.

2-3 from the scape base, capillary, 3-5 mm. long. Leaves very numerous, digitately divided to the base into 2-5 primary segments, 3-30 mm. long; primary segments ovate in outline, repeatedly dichotomously divided; ultimate segments capillary, setulose. Traps inserted in the angles between the primary segments and between subsequent dichotomies, very variable in number and size, ovoid, shortly stalked, 1-6 mm. long; mouth lateral; upper lip with 2 simple to copiously branched hairs; lower lip naked or with shorter simple hairs. Inflorescence lateral, 1·5-18 cm. high; flowers 1-3(4), distant, the lowermost sometimes very near the base of the scape; scape erect or rarely twining, filiform; scales absent; bracts basifixed, quadrate or orbicular, ± encircling the base of the pedicel, 1-3 mm. long; bracteoles absent; pedicels filiform, erect at anthesis, strongly reflexed or recurved in fruit, 2-35 mm. long. Calyx-lobes subequal, broadly ovate, 1-2 mm. long, scarcely accrescent, apex obtuse or rounded. Corolla pale to quite deep yellow, usually with brown or reddish nerves, ± densely covered externally with fine short hairs, rarely glabrous, 3-15 mm. long; upper lip broadly ovate or orbicular, apex truncate or rounded, entire or emarginate, 1·5-3 times as long as the upper calyx-lobe; lower lip orbicular to subreniform, apex ± emarginate or 2-lobed; palate raised, bigibbous; spur cylindrical, ± as long as the lower lip. Filaments linear; anther-thecae ± confluent. Ovary globose; style short or obsolete; stigma lower lip semi-orbicular, sometimes ciliate, upper very short or obsolete. Capsule globose, circumscissile, 3-4 mm. in diameter, few-many-seeded. Seeds lenticular, angular, 0·4-0·8 mm. wide, often narrowly winged on the angles; testa reticulate, cells distinct, isodiametric or ± elongate.

var. reflexa

Corolla 6-15 mm. long; lower lip longer than the spur, apex emarginate or bifid. Capsule 4-5 mm. in diameter; pedicel reflexed. Seeds 0·6-1 mm. wide.

UGANDA. Acholi District: Chua, Paranga, 12 Dec. 1935, *A. S. Thomas* 1556!; Teso District: Omunyal Swamp, 14 Sept. 1954, *Lind* 414!; Masaka District: Lake Nabugabo, 6 Oct. 1953, *Drummond & Hemsley* 4641!
KENYA. Lake Naivasha, 3 Jan. 1969, *E. Polhill* 242!
TANGANYIKA. Mafia I., 7 Aug. 1937, *Greenway* 5020!; Iringa District: Kibengu, 16 Feb. 1962, *Polhill & Paulo* 1508!; 12 km. W. of Songea, 7 Jan. 1956, *Milne-Redhead & Taylor* 8066!
DISTR. U1-4; K3; T1, 4, 6-8; Senegal to South West Africa and South Africa (Transvaal) and in Madagascar
HAB. Still or slowly flowing water in swamps and pools; sea-level to 1900 m.

SYN. *U. diploglossa* Oliv. in J.L.S. 9: 147 (1865); Stapf in F.T.A. 4 (2): 494 (1906)) Type: Angola, Huila, *Welwitsch* 271 (LISU, holo.!, BM, C, COI, G, K, P, iso.!, *U. platyptera* Stapf in F.T.A. 4 (2): 492 (1906). Type: Nigeria, Nupe, *Barter* (K. holo.!)
U. charoidea Stapf in F.T.A. 4 (2): 493 (1906). Type: Nigeria, Lokoja, *Barter* (K, holo.!)

var. parviflora *P. Taylor* in K.B. 18: 168 (1964). Type: Tanganyika, Songea, *Milne-Redhead & Taylor* 9236 (K, holo.!, B, BR, EA, LISC, SRGH, iso.!)

Corolla 3-4·5 mm. long; lower lip shorter than the spur, apex rounded, entire. Capsule ± 2 mm. in diameter; pedicel erect. Seeds ± 0·4 mm. wide.

TANGANYIKA. Songea District: 9·5 km. SW. of Songea, 26 Apr. 1956, *Milne-Redhead & Taylor* 9833!
DISTR. T8; Rhodesia and Zambia
HAB. Shallow still or slowly flowing water in bogs; 1000-1050 m.

19. **U. foliosa** *L.*, Sp. Pl.: 18 (1753); Stapf in F.T.A. 4 (2): 491 (1906); P. Taylor in F.W.T.A., ed. 2, 2: 381 (1963) & in K.B. 18: 174 (1964), excl. syn. *U. floridana*, & in Fl. Afr. Centr., Lentib.: 46, t. 14 (1972). Type: West Indies, Hispaniola, Plumier drawing

Aquatic herb. Stolons robust, oblong or elliptic in cross-section, smooth and glabrous, 1–3 mm. wide and up to several metres long; internodes 2–15 cm.; rhizoids absent. Leaves broadly ovate in outline, multiple pinnate, up to 15 cm. long, usually dimorphic, some with fewer segments and numerous traps, some with more numerous segments and few or no traps; ultimate segments capillary, minutely setulose. Traps lateral near the base of the penultimate leaf-segments, broadly ovoid, stalked, 1–2 mm. long; mouth lateral; upper lip naked or with 2 simple or sparsely branched hairs. Inflorescence erect, 7–40 cm. high; flowers 3–20, congested at initial anthesis, inflorescence-axis elongating as it matures; scape straight, relatively stout, up to 3 mm. thick, smooth and glabrous; scales absent or 1–2 just below the lowermost flower, similar to the bracts; bracts basifixed, broadly ovate or orbicular, 2–4 mm. long; bracteoles absent; pedicels at first erect, filiform, 4–10 mm. long, elongating and recurving in fruit. Calyx-lobes subequal, broadly ovate, 3–4 mm. long, connate at the base, scarcely accrescent; upper lobe apex subacute, lower 2–3-dentate. Corolla yellow, 8–15 mm. long; upper lip orbicular, about twice as long as the upper calyx-lobe, apex rounded; lower lip larger, oblate to subreniform, apex entire or emarginate; palate much raised, gibbous; spur narrowly conical, straight, ± two-thirds to ± as long as the lower lip; inner abaxial surface with 2 elliptic patches of shortly stalked glands, one on either side of the central nerve. Filaments linear; anther-thecae confluent. Ovary globose; style short but distinct; stigma lower lip orbicular, often ciliate and hispid, upper much smaller, entire or bidentate. Capsule globose, up to 8 mm. in diameter, sparsely and minutely glandular, indehiscent. Seeds 4–12, lenticular with a narrow regular wing, 2–2·5 mm. in total diameter; wing 0·2–0·4 mm. wide; testa-cells indistinct, ± isodiametric.

UGANDA. Lango District: Nabyeso, Sept. 1935, *Eggeling* 2197!; Mengo District: Lake Victoria near Kaazi, Oct. 1952, *Lind* 127!
TANGANYIKA. Buha District: Kaberi Swamp, 10 Aug. 1950, *Bullock* 3123!; Ufipa District: Lake Rukwa, *Lea*!
DISTR. U1, 4; T4; throughout tropical and subtropical Africa, Madagascar and America from Florida to Argentina
HAB. Still or slow flowing water in lakes, marshes and rivers; ± 1050–1200 m.

20. **U. australis** *R. Br.*, Prodr. Fl. Nov. Holl.: 430 (1810); P. Taylor in Fl. Afr. Centr., Lentib.: 48 (1972). Type: Australia, *R. Brown* (BM, holo.!, K, iso.!)

Aquatic herb, perennating by winter buds (turions). Stolons 2 or 3 from the base of the scape, filiform, terete, glabrous, up to 50 cm. long or more, 0·5–1·5 mm. thick; internodes 3–10 mm.; rhizoids few (2–4) from the scape-base, capillary, 1–2 cm. long. Leaves very numerous, 2-branched from the base, each branch 1–5 cm. long, ovate to lanceolate in outline, pinnately branched; pinnae alternate, repeatedly dichotomously forked; ultimate segments capillary, setulose. Traps usually numerous, lateral on the leaf-segments just above the point of bifurcation, obliquely ovoid, stalked, 1–2 mm. long; mouth lateral, oblique; upper lip with 2 slender ± branched hairs; lower lip with a variable number of simple hairs. Inflorescence erect, up to 15 cm. high; scape usually straight at anthesis but becoming very flexuous, 1–2 mm. thick, smooth and glabrous; flowers 4–6, ± congested at anthesis, inflorescence-axis elongating after anthesis; scales 1–2(3), a short distance below the lowermost flower, similar to the bracts; bracts basifixed, orbicular, ± 3 mm. long, decurrent; bracteoles absent; pedicels filiform, 1–2 cm. long, erect at anthesis, elongating and spreading or deflexing after anthesis. Calyx-lobes subequal, ovate, ± 3 mm. long; upper lobe apex rounded, hyaline; lower lobe truncate or emarginate. Corolla pale yellow, ± 15 mm. long; upper lip orbicular or broadly ovate, truncate, 2–3 times as long as the upper calyx-lobe; lower lip oblate or reniform, ± as long as the upper lip and up to twice as wide; palate raised, gibbous; spur stout, conical, slightly curved, obtuse, usually shorter than the lower lip, with shortly stalked glands on the inner adaxial surface only. Filaments filiform; anther-thecae confluent. Ovary globose, minutely lepidote; style distinct, almost as long as the ovary; stigma lower lip semi-orbicular, margin ciliate, upper almost obsolete. Capsules apparently never produced.

UGANDA. Kigezi District: Kashambya Swamp, 5 Sept. 1952, *Norman* 156a!
KENYA. Lake Naivasha, 19 Oct. 1969, *E. Polhill* 292!; Ravine District: Lake Narasha, 7 Sept. 1969, *Richardson* 33!
TANGANYIKA. Arusha District: Momela Lakes, 19 Jan. 1969, *Richards* 23782!; Ufipa District: Lake Kwela, 4 Nov. 1956, *Richards* 6857!
DISTR. U2; K3; T2, 4; Sudan, South Africa (Transvaal, Natal); mainly Europe and temperate Asia to Japan, also northern India, Ceylon, New Guinea, SE. Australia, Tasmania and New Zealand
HAB. Lakes and pools; 900–1860 m., not flowering at the lower altitudes and usually (S. of the equator) sporadic and probably introduced by migratory birds

SYN.   *U. neglecta* Lehm., Pugill. Pl. 1: 38 (1828). Type: Germany, near Hamburg, *Lehmann* (HBG, ? iso.!)
    U. incerta Kam. in E.J. 33: 111 (1902); Stapf in F.T.A. 4 (2): 496 (1906). Type: Sudan, Bahr el Ghazal, *Schweinfurth* 862 (K, NH, P, S, iso.!)
    [*U. vulgaris* sensu P. Taylor in K.B. 18: 171 (1964), *non* L.]

21. U. gibba L., Sp. Pl.: 18 (1753); P. Taylor in K.B. 18: 197 (1964) & in Fl Afr. Centr., Lentib.: 49, t. 15 (1972). Type: U.S.A., Virginia, *Clayton* (BM, holo.!)

Aquatic or subaquatic herb. Stolons usually up to ± 10, fasciculate at the base of the scape, filiform, terete, up to 20 cm. long or more, sometimes slightly inflated, up to 1 mm. thick. Rhizoids few from the base of the scape, often absent, filiform. Leaves numerous, alternate on the stolons, up to 20 mm. long; segments capillary, glabrous, simple or usually dichotomously forked from or near the base, each segment sometimes again 1–3 or rarely more times forked. Traps usually numerous, replacing one of the leaf-segments at a fork, ovoid, 1–1·5 mm. long, stalked; mouth lateral; upper lip with 2 long usually copiously branched hairs; lower lip usually with a few shorter simple hairs. Inflorescence erect, 2–35 cm. high, solitary or often several arising in succession from the fascicle of stolons; flowers (1–)2(–6); scape straight, smooth and glabrous; scales absent or more usually 1 at or above the middle of the scape, similar to the bracts; bracts basifixed, semi-orbicular, semi-amplexicaul, ± 1 mm. long; bracteoles absent; pedicels filiform, erect, (2–)6–12(–30) mm. long. Calyx-lobes subequal, orbicular to broadly ovate, 1–3 mm. long. Corolla yellow, often with brown or reddish nerves, 4–25 mm. long; upper lip orbicular, 2–3 times as long as the upper calyx-lobe, usually shallowly and obscurely 3-lobed; lower lip usually shorter and narrower than the upper, orbicular, usually entire, rarely emarginate; palate much raised and bigibbous; spur conical to narrowly cylindrical, shorter than to almost twice as long as the lower lip, apex usually bearing a few shortly stalked glands. Filaments linear; anther-thecae ± confluent. Ovary globose; style short but distinct; stigma lower lip semi-orbicular, upper short, ± obsolete. Capsule globose, 2–4 mm. in diameter, dehiscing into 2 lateral valves. Seeds rather few (± 20–40), imbricate on the smooth placenta, lenticular with a broad irregular wing, 1–1·6 mm. in total diameter, outer surface smooth to verrucose; hilum prominent; testa-cells small, irregular, indistinct.

subsp. **gibba**; P. Taylor in Mitt. Bot. Staats. München 4: 98 (1961) & in F.W.T.A., ed. 2, 2: 381 (1963) & in K.B. 18: 198 (1964) & in Fl. Afr. Centr., Lentib.: 49, t. 15 (1972)

Corolla 8–20 mm. long; upper lip 5–10 mm. wide.

UGANDA. Masaka District: Lake Nabugabo, 6 Oct. 1953, *Drummond & Hemsley* 4642!;
Mengo District: Kampala, King's Lake, 17 Sept. 1935, *Hancock & Chandler* 41!
KENYA. Naivasha District: Lake Naivasha, 6 Apr. 1961, *J. G. Williams* in *E.A.H.*
12348! & South Kinangop Plateau, Semini Swamp, 12 Mar. 1961, *J. G. Williams* in
E.A.H. 12347!
TANGANYIKA. Bukoba District: Kalema, Aug. 1931, *Haarer* 2119!; Kigoma District:
Buhamba, 2 Nov. 1953, *Ross* 1517!; Songea District: 6·5 km. W. of Songea, 28 Apr.
1956, *Milne-Redhead & Taylor* 9845!
ZANZIBAR. Pemba I., Chuaka, 10 Oct. 1929, *Vaughan* 780!
DISTR. **U**4; **K**3, 4; **T**1, 4, 6, 8; **P**; Nigeria to South Africa (Transvaal) and Madagascar,
also in America from north-eastern U.S. to Argentina
HAB. Shallow water and mud in pools and ditches; sea-level to 2460 m.

SYN. *U. obtusa* Sw., Prod. Veg. Ind. Occ.: 14 (1788); Stapf in F.T.A. 4 (2): 495 (1906)

subsp. **exoleta** (*R. Br.*) *P. Taylor* in Mitt. Bot. Staats. München 4: 101 (1961) & in
F.W.T.A., ed. 2, 2: 381 (1963) & in K.B. 18: 204 (1964) & in Fl. Afr. Centr., Lentib.: 52
(1972). Type: Australia, *R. Brown* (BM, holo.!)

Corolla 4–8 mm. long; upper lip 3–4 mm. wide.

UGANDA. Acholi District: Chua, Paranga, 12 Dec. 1935, *A. S. Thomas* 1555!; Kigezi
District: Kachwekano Farm, June 1951, *Purseglove* 3632!; Entebbe, 20 Dec. 1951,
Norman 76!
KENYA. Lake Naivasha, 5 July 1933, *Napier* 2680!; Kiambu District: Ondiri Swamp,
Feb. 1951, *Verdcourt* 428!; Lamu District: Kui I., June 1956, *Rawlins* in *E.A.H.* 2125!
TANGANYIKA. Mwanza District: Ukerewe I., *Conrads* 945!; Mbulu District: Tarangire
Swamp, 20 Nov. 1968, *Richards* 23406!; Ufipa District: Lake Sundu, 23 Nov. 1960,
Richards 13593!
ZANZIBAR. Zanzibar I., Mbiji Swamp, 8 Feb. 1929, *Greenway* 1382!; Pemba I., Pandani–
Kinazini, 19 Feb. 1929, *Greenway* 1505!
DISTR. **U**1–4; **K**3–5, 7; **T**1, 2, 4, 6, 7; **Z**; **P**; throughout tropical and subtropical Africa,
Madagascar, Portugal, tropical and subtropical Asia to Australia
HAB. Shallow water or mud in swamps, pools and ditches; sea-level to 2460 m.

SYN. *U. exoleta* R. Br., Prodr. Fl. Nov. Holl.: 430 (1810); Stapf in F.T.A. 4 (2): 4 9
(1906); F.W.T.A. 2: 234 (1931)

22. **U. cymbantha** *Oliv.* in J.L.S. 9: 147 (1865); Stapf in F.T.A. 4 (2): 494
(1906); P. Taylor in K.B. 18: 209 (1964) & in Fl. Afr. Centr., Lentib.: 52
(1972). Type: Angola, Huila, *Welwitsch* 272 (LISU, holo.!, BM, G, K, P, iso.!)

Aquatic herb. Stolons capillary, up to 10 cm. long or more, ± 0·2 mm.
thick, minutely glandular; rhizoids absent. Leaves numerous, forked from the
base into 2 equal or unequal capillary segments, 1–2·5 mm. long; internodes
1–2 mm. long. Traps numerous, inserted in the angle between the leaf-
segments or laterally on the longer leaf-segment, ovoid, stalked, 1–1·5 mm.
long; mouth lateral; upper lip with 2 capillary ± branched hairs; lower lip
with 3 shorter simple hairs. Inflorescence lateral, erect; scape capillary,
2–10 mm. high, apparently always 1-flowered; scales absent; bract basifixed,
ovoid, ± 1 mm. long; bracteoles absent; pedicel capillary, cernuous in bud
and in fruit, erect at anthesis, 1·5–4 mm. long. Calyx-lobes subequal, orbicular,
± 1 mm. long, accrescent. Corolla white or cream, 3–5 mm. long; upper lip
1·5–2 times as long as the upper calyx-lobe, semi-orbicular, apex entire or
emarginate; lower lip orbicular, base subcordate, apex rounded, entire or
emarginate; palate scarcely raised, minutely glandular; spur very short,
saccate. Filaments capillary; anther-thecae confluent. Ovary ellipsoid,
± 0·4 mm. long; ovules 2–5; style distinct, filiform, as long as the ovary;
stigma lower lip orbicular, ciliate, upper obsolete. Capsule ellipsoid, ± 1 mm.
in diameter, the wall membranous, apparently indehiscent. Seeds 2 or 3,
lenticular, narrowly and regularly winged, total diameter ± 0·7 mm.;
testa-cells indistinct, irregular, ± elongated.

UGANDA. Kampala, King's Lake, 21 July 1935, *Hancock & Chandler* 10!
DISTR. **U4**; Zaire, Mozambique, Zambia, Botswana, Angola, South Africa (Transvaal) and Madagascar
HAB. Still water in pools; 1170 m.

SYN. *Biovularia cymbantha* (Oliv.) Kam. in E.J. 33: 113 (1902)

2. GENLISEA

A. St.-Hil., Voy. Distr. Diam. 2: 428 (1833); Stapf in F.T.A. 4 (2): 497 (1906); P. Taylor in Fl. Afr. Centr., Lentib.: 53 (1972)

Rootless perennial or annual herbs of wet places. Stem short, subterranean, erect or decumbent. Leaves dimorphic, persistent at anthesis; foliage leaves petioled, entire, linear-lanceolate to spathulate or orbicular, glabrous or rarely hairy, densely or laxly rosulate from the upper part of the stem; pitcher leaves (traps) ± densely congested on the lower part of the stem and descending into the substrate and consisting of a stalk and a slender tube, cylindrical from an ellipsoid base and terminating in 2 ribbon-like helically twisted arms, the arms and tube provided internally with transverse rows of stiff inwardly directed hairs. Inflorescence terminal and arising from the leaf-rosette, racemose, bracteate; scape simple or branched above, erect, usually glandular or hispid, rarely completely glabrous, provided with ± numerous sterile bracts (scales); raceme congested to ± elongated, few–many-flowered; pedicels usually considerably longer than the bracts, erect at anthesis, erect, spreading or strongly recurved in fruit, glandular, hispid or ± glabrous. Bracts basifixed; bracteoles 2, inserted with the bract at the base of the pedicel. Calyx-lobes 5, subequal, slightly accrescent, densely glandular to hispid or glabrous. Corolla bilabiate, glandular, hispid or glabrous, blue, violet, mauve, yellow or white; upper lip ± erect, entire or 2-lobed; lower lip larger, spurred at the base; palate raised and ± gibbous, limb spreading or deflexed, ± deeply 3-lobed; spur acute or obtuse, shorter than to longer than the lower lip. Stamens 2, inserted at the base of the corolla; filaments falcate; anthers dorsifixed, ellipsoid, the thecae ± confluent. Ovary globose, glandular, hispid or glabrous; style short, indistinct; stigma bilabiate; lower lip about as broad as the ovary, semiorbicular; upper lip ± obsolete; ovules numerous, sessile on a fleshy free basal placenta, anatropous. Capsule globose, uniquely multiple-circumscissile, the lowermost line of dehiscence approximately equatorial and with 2 others between it and the persistent style. Seeds exalbuminous, numerous, ovoid, reticulate, with a prominent hilum at one end.

About 15 species, 9 in tropical America, the rest in tropical Africa, one of which occurs also in Madagascar.

Fruiting pedicel strongly recurved; ovary and calyx densely covered with gland-tipped hairs . . . . 1. *G. margaretae*
Fruiting calyx erect; ovary and calyx glabrous or hispid, the hairs not or rarely gland-tipped . . . . 2. *G. hispidula*

1. **G. margaretae** *Hutch.*, Botanist in S. Africa: 529 (1946). Type: Zambia, N. of Kasama, *Hutchinson & Gillett* 4050 (K, holo.!, BM, iso.!)

Terrestrial herb. Stem short, erect. Leaves very numerous, densely rosulate, spathulate, up to 3 cm. total length, the lamina up to 4 mm. wide. Traps numerous, up to 20 cm. total length. Inflorescence erect, simple, 20–35(–50) cm. high; scape straight, rigid, terete, 1–1·5 mm. thick, sparsely or very sparsely covered with short gland-tipped hairs below, more densely so above; flowers (4–)6–10(–20), congested; scales numerous, similar to the bracts; bracts basifixed, ovate-lanceolate, ± 2 mm. long; bracteoles similar

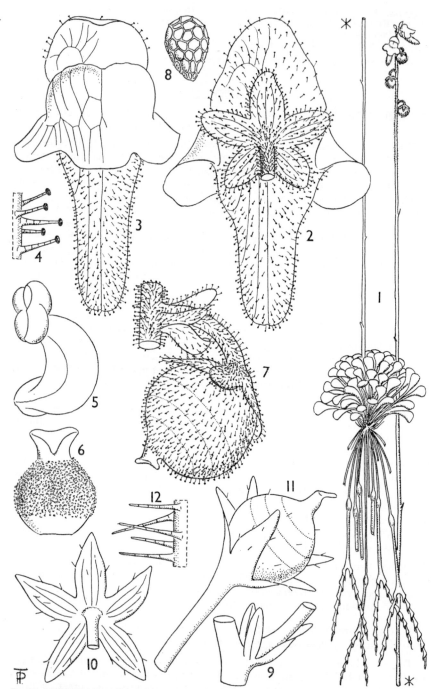

FIG. 3. *GENLISEA MARGARETAE*—**1**, plant, × 1; **2**, flower, adaxial view, × 10; **3**, corolla, abaxial view, × 10; **4**, gland-tipped hairs on pedicel, × 50; **5**, stamen, × 25; **6**, pistil, × 25; **7**, fruit, × 10; **8**, seed, × 50. *G. HISPIDULA* subsp. *SUBGLABRA*—**9**, base of pedicel with bracts and bracteoles, × 8; **10**, calyx × 8; **11**, fruit, × 18; **12**, hairs on spur, × 60. 1–8, from *Milne-Redhead & Taylor* 10013A; 9–12, from *Milne-Redhead & Taylor* 8009.

but shorter; pedicels erect at anthesis, 1–3 mm. long, elongating and strongly recurving in fruit, ± densely covered with short gland-tipped hairs. Calyx-lobes oblong-ovate, obtuse, subequal, ± 2 mm. long, 3-nerved, ± densely covered externally with short gland-tipped hairs. Corolla mauve or purple, 7–10 mm. long, covered externally with short gland-tipped hairs; upper lip ovate-oblong, apex truncate, about twice as long as the calyx-lobes; lower lip longer and broader, rather shallowly 3-lobed; palate slightly raised; spur cylindrical from a conical base, 2–3 times as long as the lower lip. Filaments falcate, ± 1 mm. long; anther-thecae subdistinct. Ovary globose, densely covered with short gland-tipped hairs; style short; stigma lower lip semi-orbicular, the upper lip obsolete. Capsule globose, 2·5–3 mm. in diameter, densely covered with short gland-tipped hairs. Seeds numerous, ovoid, 0·3–0·4 mm. long, reticulate; testa cells ± isodiametric.

TANGANYIKA. Songea District: about 6·5 km. E. of Gumbiro, 29 June 1956, *Milne-Redhead & Taylor* 10013A!
DISTR. **T8**; Zambia, Madagascar
HAB. Permanently wet bogs; 855 m.

SYN. *G. recurva* Bosser, Le Naturaliste Malgache 10: 23 (1958). Type: Madagascar, *Bosser* 11707 (P, holo.!)

2. **G. hispidula** *Stapf* in Fl. Cap. 4 (2): 437 (1904) & in F.T.A. 4 (2): 498 (1906); P. Taylor in F.W.T.A., ed. 2, 2: 375 (1963) & in Fl. Afr. Centr., Lentib.: 58 (1972). Type: South Africa, near Pretoria, *Kirk* 36 (K, lecto.!)

Terrestrial herb. Stem short, erect. Leaves usually few, laxly rosulate, spathulate, usually long-petiolate, total length (1–)2–4(–5) cm., the lamina up to 9 mm. wide. Traps about as numerous as the leaves. Inflorescence erect, simple or sparsely branched above, (6–)10–25(–50) cm. high; scape flexuous, terete, glabrous above, usually sparsely hispidulous at or near the base only; flowers (1–)3–10(–15); scales few, similar to the bracts; bracts ovate-lanceolate to linear-lanceolate, acute or acuminate, 2–4 mm. long, glabrous to ± hispidulous; bracteoles similar but narrower; pedicels erect, (2–)5–15(–25) mm. long at anthesis, usually elongating in fruit, glabrous or ± hispidulous, the hairs rarely gland-tipped. Calyx-lobes subequal, ovate-lanceolate to lanceolate, acute or acuminate, 2–4 mm. long, sparsely to densely hispidulous or rarely glabrous. Corolla violet, blue, mauve or pink with a greenish or yellowish spur, (6–)10–12(–15) mm. long; upper lip ovate, about twice as long as the calyx-lobes; lower lip larger, broader than long, 3-lobed; palate raised, gibbous; spur cylindrical, obtuse, 1½–2 times as long as the lower lip, ± hispidulous, the hairs rarely gland-tipped. Filaments falcate; anther-thecae subdistinct. Ovary ovoid, glabrous to densely hispidulous; style short, indistinct; stigma lower lip semiorbicular, upper obsolete. Capsule globose to broadly ovoid, 3–4 mm. long, glabrous to hispidulous. Seeds numerous, ovoid, 0·4–0·5 mm. long, conspicuously reticulate; testa cells ± isodiametric.

subsp. **hispidula**

Ovary and capsule ± densely hispid, at least in the upper part.

KENYA. Cherangani Hills, Kaibwibich, Aug. 1968, *Thulin & Tidigs* 182!
TANGANYIKA. Iringa District: 13 km. S. of Dabaga, 22 Feb. 1962, *Polhill & Paulo* 1572!; Njombe District: Madehani, 3 Dec. 1913, *Stolz* 2319!
DISTR. **K3**; **T7**; Nigeria, Cameroun, Central African Republic, south tropical and South Africa
HAB. Wet soil or shallow water in bogs and marshes; 2100–2700 m.

subsp. **subglabra** (*Stapf*) *P. Taylor* in K.B. 26: 444 (1972) & in Fl. Afr. Centr., Lentib.: 58, t. 17 (1972). Type: Zambia, Lake Tanganyika, Fwambo, *Nutt* (K, lecto.!)

Ovary and capsule glabrous or almost so.

TANGANYIKA. Songea District: Ulamboni valley, 11 km. W. of Songea, 31 Dec. 1955, *Milne-Redhead & Taylor* 8009 !
DISTR. **T8**; Burundi, Zaire (Katanga), Zambia, Malawi
HAB. Marshes and swampy grassland; 960–1500 m.

SYN. *G. subglabra* Stapf in F.T.A. 4 (2): 498 (1906)

INDEX TO LENTIBULARIACEAE

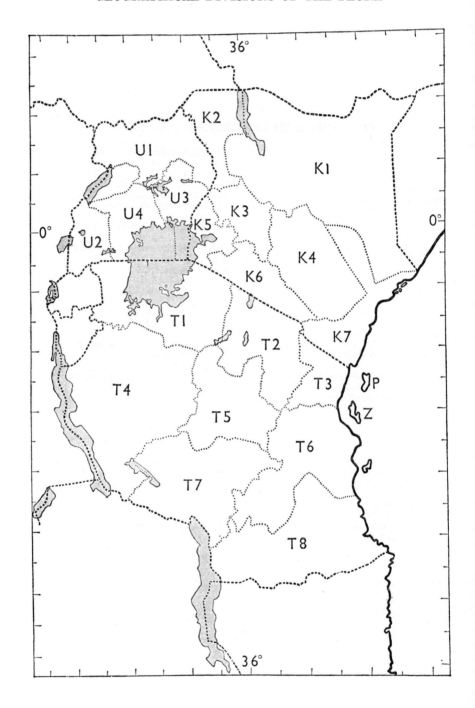